WRITTEN BY

LAURENCE OTTEINHEIMER-MACQUET, FLORE D'ARFEUILLE, RAPHAELLE BRICE,
KATIA FORTIER, CLAIRE JOBIN, MARIE-PIERRE KLUT, ODILE LIMOUSIN,
JEAN-PIERRE REYMOND, ALINE RIQUIER

COVER DESIGN BY
STEPHANIE BLUMENTHAL

TRANSLATED AND ADAPTED BY
LINDA BUTURIAN AND ROSEMARY WALLNER

PUBLISHED BY CREATIVE EDUCATION
123 South Broad Street, Mankato, Minnesota 56001
Creative Education is an imprint of The Creative Company

Library of Congress Cataloging-in-Publication Data
[De quoi sont faits les objets familiers? English]
How things are made / by L. Otteinheimer-Macquet et al. ;
[translated and adapted by Linda Buturian and Rosemary Wallner].
(Creative discoveries)
Includes index.
Summary: Describes the origins of and materials used in making a variety of objects
including paint, wool, glass, and plastics.
ISBN: 0-88682-955-0
1. Manufactures—Miscellanea—Juvenile literature. [1. Manufactures.]
I. Ottenheimer-Macquet, Laurence. II. Title III. Series
TS146.D413   1999
670—dc21   97-27526

2 4 6 8 9 7 5 3 1

HOW THINGS ARE MADE

CONTENTS

CREATIVE EDUCATION

Take a walk down your street and notice all the things that are made of rock. You might see brick houses, cement steps, sidewalks, and the street itself. Rock, or stone, is a good building material because it stands up to hot, cold, and wet weather.

People have always used stone. Prehistoric people cut and polished stones to use as tools and weapons. Later on, they realized that the most useful thing about stone is that they could build with it. A shelter made of stone kept them warm and dry, protected them from the wind, and lasted a long time. The ancient Egyptians used huge blocks of stone to build their pyramids, which have stood in the desert for thousands of years.

Stones, or rocks, form Earth's crust—the outermost layer of the planet. Most stones are hard solids, but some are different. Clay is soft enough to mold into shapes. Sand is stone worn down into tiny grains.

A gravel road

Earth: The original recycler

Rocks appear fixed and solid, but every day they are exposed to physical conditions that, over time, change them. This exposure is called the rock cycle. It's the slowest of Earth's processes and, over millions of years, recycles materials. This recycling process creates deposits of mineral resources we depend on every day. Geologists group rocks into three main families according to how the rocks were formed: igneous, sedimentary, and metamorphic.

Iron pyrites

Fluorspar

Rock salt

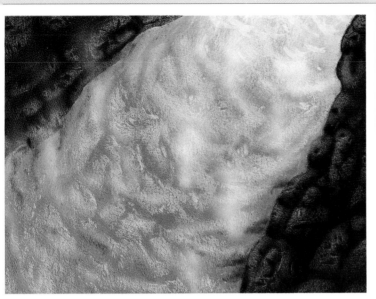

Liquid rock, called lava, streams from a volcano's crater like a river of fire. It cools in the air and hardens into volcanic rock.

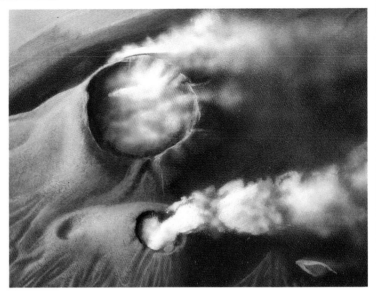

An active volcano has smoke billowing out of its cones and small amounts of lava flowing down its sides.

Igneous rocks are formed beneath and on volcanoes. The Earth's center is so hot that rocks melt and turn into a liquid called magma. A volcano erupts when magma forces its way through a crack in the Earth's crust.

Obsidian is a hard volcanic rock that looks like dark glass. **Basalt,** another volcanic rock, forms as lava cools. Devil's Postpile National Monument is a group of basalt blocks in the Sierra Nevada mountains of California. The constant flow of the river wore away some of the rocks, making them look like huge, uneven stairs.

Volcanic rocks: obsidian, basalt, and vesicular basalt containing air bubbles

Once the magma reaches the Earth's surface, it is called lava. When lava meets the air, it cools down and becomes solid rock that is gray or black. The world's most active volcano is Kilauea in Hawaii. Kilauea has been erupting almost nonstop since 1983. **Pumice,** a rough volcanic rock that you can use to clean your hands, is so light that it floats in water. Pumice is actually solidified volcanic foam full of gas bubbles that were trapped in it as it cooled.

Some rocks built up from the remains of dead plants and animals.

Shellfish, like this nautilus, die and sink to the seabed.

Some rocks form slowly over millions of years, gradually accumulating objects such as bits of shell and animal bones. These are **sedimentary rocks.** On the stone faces of the Grand Canyon in Arizona, you can see the layers of different rocks stacked one on top of another. A layer of white rock found in the Grand Canyon is kaibab limestone, formed about 250 million years ago. **Limestone** is a common sedimentary rock made up of shells, algae, and pieces of coral that drifted down to the seabed long, long ago.

Millions of years after they have died, shellfish have become fossils preserved in the sedimentary rock.

People use electricity to switch on a computer, TV, or light—but where does electricity come from? In the United States, coal is one of the three main sources of energy for electricity.

How was coal formed? Millions of years ago, large forests and jungles of giant ferns grew in swamps and shallow lagoons. As the plants and trees died, they broke down into hydrocarbons. Gradually, layers of mud and rock compressed these remains, squeezing out all the water and leaving deposits of the hard black substance we call coal. Many different types of coal exist. The best quality for burning is anthracite, from the layers that have been compressed the

In the early 19th century, adults and children worked in dreadful conditions deep inside coal mines.

longest. It is hard and shiny. Lignite (brown coal) is softer and newer, and gives less heat when it burns.

People have mined coal for centuries. To get coal near the Earth's surface, miners use a method called strip mining. The miners dig a quarry and clear away the soil; then they use machines to cut the coal out of the ground.

When the coal seams are farther down, miners dig an underground mine with a vertical shaft—which sometimes goes down more than 3,280 feet (1,000 m) below the Earth's surface. Lifts carry the coal to the surface where it is sorted into different grades, washed, and broken up.

While coal is an important **energy source** for heating and electricity, extracting it from the ground is hard on the environment, and burning coal pollutes the air. Scientists and engineers are looking for better ways to reclaim land mined for coal and reduce air pollutants.

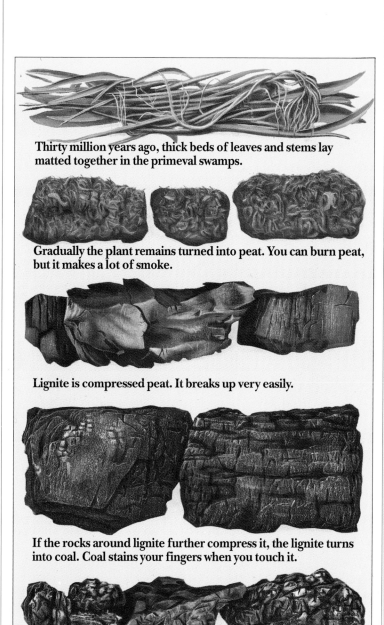

Thirty million years ago, thick beds of leaves and stems lay matted together in the primeval swamps.

Gradually the plant remains turned into peat. You can burn peat, but it makes a lot of smoke.

Lignite is compressed peat. It breaks up very easily.

If the rocks around lignite further compress it, the lignite turns into coal. Coal stains your fingers when you touch it.

The last stage is anthracite, the best coal of all. It burns with little smoke and gives off lots of heat.

People create buildings out of stone.

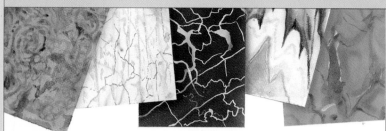

Since the days of the ancient Greeks, people have prized marble for its beauty. Marble comes in almost any color and can be polished until it shines.

Metamorphic rocks are named for the Greek word for "change of form." Violent upheavals at the Earth's center pushed and cracked the crust—and created mountains. As the rocks folded over each other, intense heat and pressure changed their form. **Marble** is limestone hardened by heat and pressure. Because people can easily cut and polish marble, they have used it for centuries to build and decorate magnificent temples, churches, and palaces. Artists have used blocks of marble to carve many of the world's greatest sculptures.

A marble quarry. The most famous marble in the world comes from Carrara in Italy where Michelangelo bought his stone.

Stone is used for building. Because stone is heavy and difficult to move, people usually built their houses using nearby rock and stone. Limestone, granite, and sandstone are the main types of stone suitable for building.

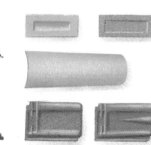

The color of bricks and tiles depends on the type of clay from which they are made.

When mud tightly compresses it forms slate. **Slate** splits easily into thin sheets, and builders often use it for covering roofs. At school, children used to write on flat pieces of slate with chalk—instead of using pencil and paper. Builders use stone to create bricks and tiles. Wet clay is cut to the right shape, and then fired in a kiln until it is hard.

The Taj Mahal in India is made entirely of marble.

Concrete is made from cement, gravel, and water.

Concrete, rather than wood or stone, is the main material in most modern buildings. Concrete is so strong that builders use it to create skyscrapers that hold hundreds of apartments or offices. Concrete contains cement, a substance which is made by heating and crushing clay, limestone, and sand.

Sand is rock that wind and water have ground down into tiny particles. Workers extract sand from rivers and dig it from quarries. They can also take sand from the sea, but then they must wash it to get rid of the salt. Buildings and pipes can be cleaned by sandblasting.

Gems are rare and precious stones.

Rock crystal

Some people believe that rocks have mysterious powers. The Aboriginal people of Australia believe the sky is a dome made of rock crystal, which they consider a magic stone. Some Native American tribes believe that crystals and other stones contain spiritual powers.

Ruby, sapphire, emerald, and diamond are the four types of **precious stone.** The diamond is the brightest, purest, and hardest of these stones; it is Earth's hardest substance. A diamond is so hard that the only thing that can cut it is another diamond. Africa produces most of the world's diamonds.

Rough pink diamond

Diamonds come in many colors. An impurity in the crystal gives the stone its color.

Jewelers cut gemstones into symmetrical shapes so that they reflect light. Skillful cutting and polishing brings out all their color and sparkle.

Ruby   **Sapphire**   **Emerald**

Small amounts of rare precious gems have been discovered in the United States. Diamonds were mined in Arkansas; emeralds, rubies, and sapphires are still mined in North Carolina.

Precious stones are rare, but you may be able to find **semiprecious stones,** which are more common. Their value depends on their transparency and purity. Transparent stones, like garnet and topaz, sparkle brilliantly. Translucent stones, like amber and agate, have a softer, milkier look. Opaque stones, such as turquoise and jet, look like beautiful pebbles. You can't see through them, but the various shades of color produce delicate patterns. All of these stones are natural minerals.

Stones can be ground into powder for paints.

Zircon Tourmaline Garnet Topaz Amethyst

Agate Chrysoprase Amazonstone Amber

Lapis lazuli Turquoise Malachite Jet

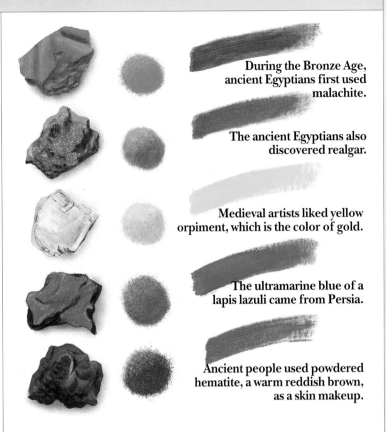

During the Bronze Age, ancient Egyptians first used malachite.

The ancient Egyptians also discovered realgar.

Medieval artists liked yellow orpiment, which is the color of gold.

The ultramarine blue of a lapis lazuli came from Persia.

Ancient people used powdered hematite, a warm reddish brown, as a skin makeup.

What gives stones their color? Stones take on the color of the chemicals, often metal oxides, in the rocks around them. For example, rubies are red because they contain chromium.

Some Native Americans painted their faces in different ways and colors, according to the occasion (for example, for a battle, a hunt, or a celebration). They ground stone to a fine powder to make the paints. Even in prehistoric times people knew how to make paints from crushed rocks. They used the paints to draw pictures on the walls of caves.

Prehistoric artists used a mixture of brown and green clays, chalk, and charcoal. With these they drew the animals they saw when they went hunting: reindeer, bison, and their own hunting dogs.

The discovery of metals

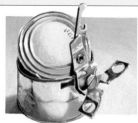

Look around you at all the objects made of metal. Cars, cans, zippers, forks, and knives are just a few. Metals are hidden underground. Sometimes they are found as lumps of **metal,** but more often they are mixed with rock in the form of ore.

How can you recognize a metal? Metal is shiny and opaque—you can't see through it. Although metal feels cold to the touch, it conducts heat as well as electricity. Mercury, used in thermometers, is the only metal that is not solid at room temperature. Some metals, like gold and silver, are rare and precious, while others are more common. Iron is the most common metal of all.

You can even find metal in food! Iron, found in red meat, fish, and peas, helps to

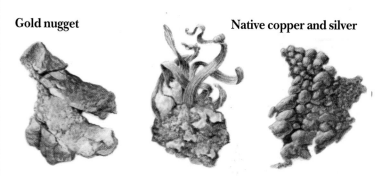

Gold nugget   Native copper and silver

People sometimes find lumps of pure metal, called nuggets or native metals, in the ground.

make you strong. Calcium, found in dairy products, builds your bones and teeth. Phosphorus, also found in dairy foods, is believed to be good for your memory.

Using metals helped man to develop and progress.

The discovery of fire opened up new possibilities: people found they could separate, or extract, metals from their ores. When people heated the ore with charcoal, the metal was released, leaving behind the rock. They could pour off the molten metal into molds to make all sorts of useful things.

This sword from northern Europe was cast in a mold.

Using heat to extract metal from ore is how our ancestors first made metal tools and weapons, such as saws, axes, and swords. About 4,000 years ago, people began to mix two metals together. Copper and tin were melted together to form bronze, a much tougher metal than copper or tin on their own.

Iron key

Greek helmet made of bronze

A mixture of different metals is called an alloy. Using bronze, people made strong weapons, tools, and sculptures. The Bronze Age had begun.

Today, **iron** is one of the metals we use most, especially in the form of steel. In the ground, iron ore is plentiful, but to obtain the metal the rock has to be heated to more than 1,832° Fahrenheit (1,000° C).

Before people could create this much heat, they used iron from meteorites that had fallen to Earth from space.

The Iron Age

The Hittites, who lived in the Middle East about 3,000 years ago, learned to build furnaces in the ground. These burned hot enough to melt iron, which soon replaced bronze because it is lighter, stronger, and easier to work with.

This meteorite was found in Greenland. For centuries, the Inuit people have used iron from meteorites to make their weapons.

People still used bronze for some things. It was better for making jewelry, because it could be shaped in a mold. Iron was used for weapons and heavy tools.

Bellows keep the furnace white-hot. The waste material runs out through a hole at the front, while the metal sinks to the bottom of the furnace.

Smiths forged red-hot metals into different shapes.

Blacksmiths shaped the hot iron with a hammer. They used pincers to hold the metal on the anvil while they worked it.

A smith is a metalworker. When iron had been extracted from its ore, smiths used to beat it hard to get rid of the impurities. Then they heated it in a forge fire. When the iron was red-hot, they worked it with a hammer on an anvil. The flat top and rounded point of the anvil helped the smiths to shape the metal in different ways. Assistants used a pair of bellows to keep the fire hot. Sometimes the smiths plunged the hot metal into water, hardening it a little before working it.

Gold ingot

A Lydian coin (600 B.C.)

Early medieval gold ring

Gold is the most precious of metals, and the most ductile of all metals, which means that a single ounce of gold can be stretched into a wire more than five miles (8 km) long! Gold is an excellent conductor of heat and never rusts. Early coins were made of gold or silver. King Croesus of Lydia was one of the first rulers to put his seal on a coin.

In the Middle Ages, armorers made armor for knights. They signed their work with a stamp or engraving.

Blacksmiths didn't just shoe horses. They often did the sort of work that veterinarians do today.

From needles to clocks . . . the ancient craft of metal working

Anything made of gold or silver must be stamped with a hallmark. This indicates when and where it was made. Artisans may also sign a piece with their own mark.

Locksmiths in the 19th century made strongboxes and steel beams as well as locks and keys.

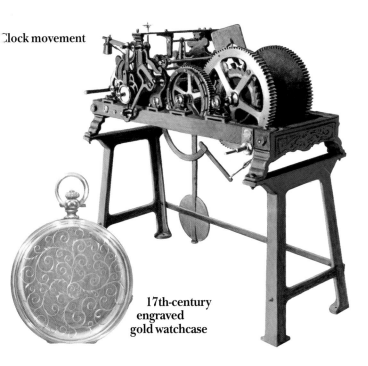

Clock movement

17th-century engraved gold watchcase

Cutlers in the 17th century used a whetstone to sharpen the knives they made.

Jewelers hallmark their work, and may sign it as well. They use precious metals such as gold, silver, and platinum, shaping the metals in molds or hammering them into a shape.

Clockmakers use precious metals to create precise cogwheels to turn the mechanisms of their clocks.

Nailmakers in the 19th century made all sorts of nails. This nailmaker's dog worked the bellows by running in a treadwheel.

Coppersmiths make pots and cauldrons by hammering the soft metal around a wooden post.

Indian metal workers engraved and chiseled designs on the dishes and vases they made.

The first mines were open cast mines, or quarries. As the need for metals grew, miners dug deeper and deeper into the earth. The work, underground in near darkness, was dirty and dangerous—tunnels could cave in. Or a miner might

inhale too much of the trapped dust and develop black lung disease, which was often fatal.

Miners went down into the mine early in the morning and stayed underground all day long.

Today, more than 100,000 men and women work in mines. It is still dangerous work. A mine must always have at least two vertical shafts: one to allow fresh air in, and another to let stale air out. Without this ventilation system, the miners would not be able to breathe. Trucks on a conveyor belt carry the ore to the shaft. Then the ore is raised to the surface on an elevator.

Today, miners wear protective helmets with headlights to light their way.

Beverage cans, foil, and airplanes are made with aluminum. **Aluminum** was first isolated a bit more than 100 years ago; it's now the second most used metal in the world after iron. Aluminum is extracted from an ore called bauxite. Smelting aluminum became possible when a new fuel called coke was available. A coke furnace can get much hotter than one that burns wood or coal.

Pure aluminum is too soft to use, but it can be mixed with other metals to form lightweight alloys that are especially well-suited to making aircraft bodies. Today

Bellows were continually improved to make the fires give out more heat.

In the Middle Ages, some smiths used water-driven hammers to help work the metal.

aluminum is smelted by a process that uses huge amounts of electricity. Recycling aluminum reduces energy consumed and pollution produced.

Metallurgy is the science of metals. During the last two hundred years, miners have discovered several new metals, including nickel, zinc, titanium, and uranium.

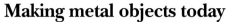

The age of steel began in the 19th century. At that time, most of the big steelworks were built. Steel is a mixture of iron and carbon—together they make an alloy that is strong and long lasting. A modern furnace used for mixing metals is different from those used before the 19th century. Today's furnaces have a large well at the bottom to hold the molten metal; the furnace can reach a temperature of 2,912° Fahrenheit (1,600°C).

Making metal objects today

First the metal is separated from the ore, and then it is refined, or made as pure as possible. After that it might be mixed with other minerals to make an alloy. Finally, it is shaped. It might be flattened and run through rollers to make girders and rails or molded into shape for the bodywork of cars.

What can we do with steel?

A steelworks

By adding small quantities of other metals to the original alloy, steel can have different qualities. Thanks to steel, people can build skyscrapers. But architects have to be careful to allow for the way metals behave at different temperatures. On a cold day the Eiffel Tower in Paris is four inches (10 cm) shorter than it is on a hot day!

Metals need protection. Water and salt make iron and steel rust, so they must be painted or covered with other metals that don't rust. Stainless steel, for example, contains chromium and nickel, which stop rust. Steel is an expensive product, both to make and to buy. Some countries have a wealth of steel—others must import it or rely on other resources.

Semi-trucks are made of steel.

These massive steel girders form the framework of a skyscraper.

The Earth's resources won't last forever.

We can reuse the metal found in things such as old cars, bicycles, and factory waste. Scrap-metal workers, people who collect these used metal objects, break up the objects, melt them down, and recast them. Due to this process, almost half of the platinum and lead used today has been recycled.

Although metal is useful, some metal waste can be dangerous. Uranium remains harmful for thousands of years. Fish die from swallowing poisonous metals that factories have dumped into rivers. Birds and humans who eat the contaminated fish may become ill and even die. Many modern factories control their waste through filtering and recycling systems. But we must all be careful of what we throw away and where we throw it.

Metals such as nickel and copper have not yet run out, but they are becoming more and more difficult and expensive to mine on land. Extracting and processing minerals requires vast amounts of energy and is a major contributor to land, air, and water pollution.

We have hardly begun to tap the riches under the sea, but we must do it carefully. Undersea miners use remote-controlled equipment to mine the ocean floor. This equipment plucks up potato-sized rocks of manganese—a metal used in steel alloys. Scientists want to experiment with refining metals in space, outside Earth's atmosphere. Without the pull of gravity to affect them, metals mined in space may contain new and exciting properties.

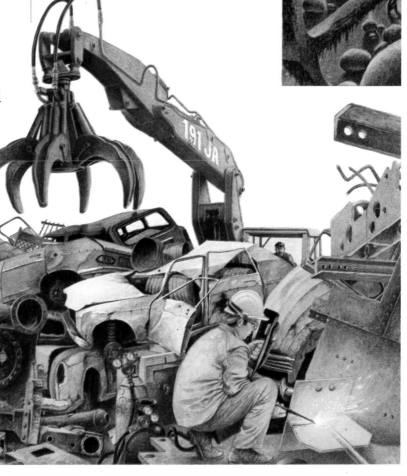

The ancient Egyptians knew the secrets of glass . . .

 Four thousand years ago, children in ancient Egypt played with glass marbles. But ancient glass was opaque, unlike the transparent substance we use today for windowpanes, drinking glasses, eyeglasses, and light bulbs. Today factories use glass to make tennis rackets, wind-surfing boards, boats, car bodies, helicopter blades, and materials that resist fire.

What is glass made from? Glass starts out as a hot, soft paste, made by melting together various ingredients. The paste is shaped while it's hot, and hardens as it cools. When the Egyptians wanted to make a bowl, they first made a sand mold and then wound a spiral of glass paste around the mold. They fired the glass in a kiln and then shook out the sand. This glass did not look like ours—it was opaque and sometimes brightly colored like china. The Romans discovered how to make glass transparent by adding silica, which is found in sand, to the mixture.

Why do we use glass? Glass is an insulator. The glass found in windows keeps out noise and cold. If you put a thin coating of silver on the back, glass becomes a mirror and reflects light. If you hold a mirror so that it catches sunlight, you can light up a dark area.

Egyptian glassmakers (left). In a heated greenhouse or conservatory made of glass, plants thrive even when the outside air reaches freezing temperatures.

Glass was still a luxury product in the 17th century.
The Hall of Mirrors in the palace at Versailles in Paris was the pride and joy of King Louis XIV.

Glass is easy to keep clean and does not absorb smells. This is why many companies store their products, such as jelly, honey, milk, and pickles, in glass containers. These companies sterilize the bottles and jars to kill any bacteria. Scientists also use glass test tubes for experiments because glass doesn't react with most chemicals.

Be careful—it's made of glass! Some glass breaks easily, and you can be cut badly with slivers of glass. But researchers have invented ways of strengthening glass. Some glass products can now be so strong they're practically unbreakable.

   Wired glass is made by pressing an iron mesh into softened glass. This type of glass is used in some public build-ings. If there's a fire in the building, the glass melts and sticks to the wire rather than breaking and falling on someone. Blasts of cold air quickly cool toughened glass, which is used for kitchen jugs and bowls. Vehicle windshields are made with laminated glass, which has a layer of plastic sandwiched between two sheets of glass. Glass can also be made bullet-proof and shatter-proof.

Glass is difficult to cut because it is hard and brittle. Most glass can be cut with a blade of tungsten, a very hard metal. The most delicate glass still has to be cut with a sharp-edged diamond, which is Earth's hardest substance.

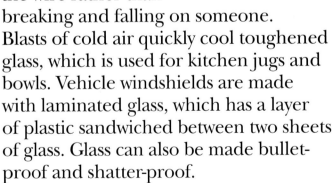

Making glass

One needs sand from a quarry, lime (a white chalky substance), and soda (a dangerous chemical because it reacts violently to air and water). First, all the ingredients are tipped into a large clay pot called a crucible and heated in a furnace. The temperature has to reach 2,732° Fahrenheit (1,500°C). At this temperature, the ingredients fuse together to make a thick liquid, like an extremely hot syrup. Glassblowers use special tools to create different shapes from the thick liquid.

The first glassworks used furnaces heated by wood. They were often built close to forests, where fuel was easy to obtain.

A glassworks

. . . by skilled glassblowers.

Glassblowers collect a blob of molten glass on the end of a blowpipe, which is a hollow metal rod. They shape the glass by rolling it along the smooth edge of a special table called a marver. They blow as hard as they can down the pipe, so the soft glass balloons into a bulb shape known as a parison. They must work quickly, before the glass cools and becomes too stiff to shape. Many artists and craftspeople still practice the fine art of glassblowing.

Some of the tools used in glassblowing: hollow wooden molds, scissors, and tongs.

To create a wineglass, glassblowers mold the parison to make the top of the glass. Then they blow it until it is the right shape and size.

They use scissors to cut off leftover glass—but keep just enough glass to make the stem, which they pull down with tongs.

Assistants bring another blob of glass paste to make the foot of the glass.

Glassblowers smooth the foot against the wooden mold and then free the glass from the pipe by giving it a sharp tap.

They use a blowtorch to soften the rim and round off the edge.

Finally, the glass is annealed. It is reheated to 932° Fahrenheit (500°C) to make it less brittle.

Stained glass
Strips of lead join together pieces of colored glass to create stained glass windows. Artists have created windows in this way since the Middle Ages.

Crystal
When creating crystal, the most transparent kind of glass, lead is added to the usual mixture. Crystal makers engrave flowers, stars, and other designs that glitter like tiny mirrors. Making and decorating crystal is slow, skilled work, so the completed objects are often expensive.

Medieval glass guildsmen installing a stained glass window

We use glass every day.

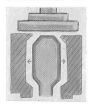

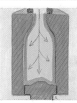

Today most of the glass we use is made in factories. Factory machines create many of today's bottles. First, a dispenser drops a blob of molten glass into a mold. Air is puffed into the glass to make a hollow area and shape the bottle. The mold is turned upside down and cooled by a jet of cold air. After the glass cools completely, it is removed from the mold. Glass is 100 percent recyclable—the process of recycling glass is efficient and produces no unwanted waste.

Windowpanes are made by floating a layer of molten glass on a bed of smooth, shiny, liquid tin. This produces a flat sheet of glass.

Lenses made of ground and polished glass are used for eyeglasses, magnifying glasses, microscopes, and telescopes.

Fiberglass is made by forcing molten glass through thousands of tiny holes in the bottom of a container, forming fine threads. If you shine a light at one end of a bundle of glass fibers, the light waves will travel along the fibers to the other end. **Fiber optics** transmit TV images, electronic information, and telephone com-

Bundles of optical fibers have replaced the old system of transmitting TV images by cable.

munications through long bundles of glass fibers. These glass fibers also can be woven into cloth or matted together to make insulating material. The fibers can be mixed with synthetic resin to make car bodies and boat hulls, which will not rust like metal hulls. Tennis rackets and fishing rods made of fiberglass are much lighter than ones made of wood.

How many things made of glass and fiberglass can you see in this picture?

Early people who lived near forests discovered the many uses of wood. It was one of the first materials prehistoric people used. They created wooden spears for hunting, fishing, and fighting other tribes. Later, when people started to work with metal, they used wood fires to melt and shape the metal. They cut down trees to build houses and boats and make all sorts of things, from tools to musical instruments. Today, we use so much wood that trees are grown especially for our use.

The earliest axes had wooden handles.

Heating the points of wooden spears and arrows made them harder and stronger.

Even so, huge areas of natural forest are cut down—thousands of acres every year—for use as lumber.

What is a tree's wood made of?

Many long chains of tiny cells, called fibers, join end to end to make up a piece of wood. Sap travels up from the roots through the fibers to feed the tree. If you break off a twig, you'll see the ends of the fibers in the wood. Where someone saws through a big branch or tree trunk, you can see the growth rings, one for each year of the tree's life. Count the rings, and you'll know the tree's age.

Wood is hard and solid but can be cut easily to shape. You can use it for some things, like walking sticks, without even shaping it. Prehistoric people would take a branch from a tree to use as a club.

Wood is an excellent source of fuel. Once people discovered how to light a fire with scraps of twigs and feed the flames with larger pieces of wood, life became more comfortable. Fire kept early people warm, and cooking their food over it was better than having to eat everything raw.

Early people took to the seas on wooden rafts.

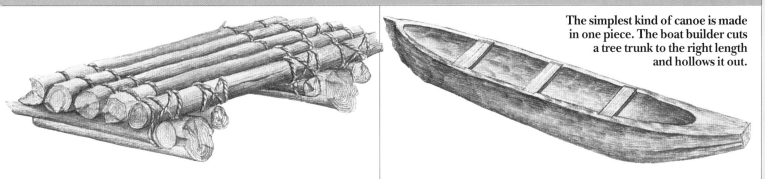

The simplest kind of canoe is made in one piece. The boat builder cuts a tree trunk to the right length and hollows it out.

Wood is light, and it floats. The first boats were just tree trunks tied together to form rafts. Prehistoric people sometimes traveled long distances on these rafts.

Wood can protect us from rain, wind, and cold. You can build a simple shelter by weaving branches together. Some tribes of Pygmies in Africa still build their huts around a framework of branches stuck into the ground.

People used tree trunks to transport objects on land before wheels were invented. Slaves in ancient Egypt used wooden rollers to move enormous blocks of stone, which were used to build pyramids and temples. Without these tree trunks, the slaves could not have hauled the stones to the building site.

The birth of technology

Once people discovered how to make tools out of hard substances like stone and metal, they began to work more with wood. With axes and hatchets they could cut down whole trees, which they could use to build homes. The first buildings were log cabins, made of tree trunks fixed one on top of another. Then people learned to make saws, so they could cut tree trunks into planks and use timber much more economically.

Craftspeople used to make all sorts of everyday objects from wood. Carpenters built the frames of houses and cut floorboards. Joiners and cabinetmakers made furniture. Wheelwrights made wooden wheels for carts and carriages, while the cartwright built the body of the cart. Many people were needed to create wooden casks for different purposes: the wet cooper made bushels (containers) for wheat and other grains, and the white cooper made churns and other dairy equipment.

The turner used a lathe to make round wooden objects such as buttons, bobbins, pulleys, handles for pistols, and tools.

2000 B.C.

1500 B.C.

1000 B.C.

Norway, A.D. 850

Europe, 1850

Wheel with tire, about 1900

Today craftspeople still create useful objects, although most are produced in factories and made of metal or plastic, not wood. Wood is beautiful and gets more beautiful as it ages—which is one reason why people collect wooden antiques.

If you cut a slice from a tree trunk and use it as a wheel, it will break easily. People realized that they could make a wheel stronger if it were built in sections, and that it would be lighter if it had spokes instead of a solid center.

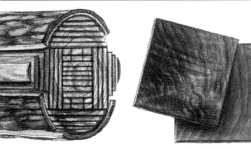

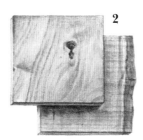

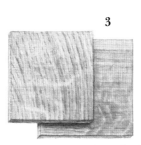

Different types of wood are cut up in different ways and used for various purposes including beams, planks, and shingles.

Half-timbered houses are buildings built on a wooden framework. To build this type of house, local carpenters would cut large vertical and horizontal beams and

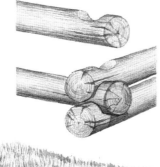

number each one so that they would know where it belonged. (If you look carefully in older houses, you might be able to see these numbers carved on the ends of the beams.) When the framework of the house was ready, the gaps were filled. In some countries, carpenters used wooden planks; in other countries they used thin pieces of wood held together with clay and plaster. Carpenters need to know all about the different types of wood, so they can choose the best type for each job.

Every time the Mongolian nomads of Asia move on with their animals, they take down their tents, or yurts, (above) which have a light wooden frame. Canadian log cabin (right)

Deciduous trees like walnut (1) and oak (2), usually give hardwood; softwood mostly comes from conifers.

Many types of wood change color; usually wood becomes darker when it's exposed to the air. Cherry wood (3) becomes a dark red color as it gets older.

Iroko (4), like teak, is a tropical wood that doesn't rot, even if it is left out in the rain.

There is nothing quite like wood . . .

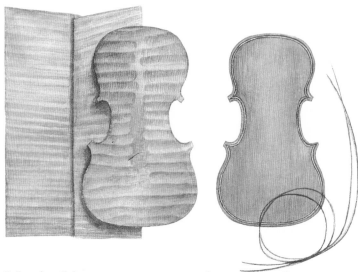

To build a wooden ship, you needed perfect timber. Tall pine or larch trunks were used to make a ship's masts. The framework of the hull, together with the forecastle and stern, were built of curved pieces of oak—which had to be grown especially for this purpose and were expensive. The planks for the keel had to be absolutely straight. They came from trees planted so close together that they were forced to grow straight up. In the 18th century, building a seventy-four-gun frigate meant cutting down more than 3,700 trees.

Good trees for ship building

The front of a violin is made of softwood, like pine or spruce. The back is made of hardwood, usually maple.

Musical instruments can be made from trees. Take a tree stump, hollow it out, pull a skin tightly over it, and you have a drum. Tom-toms are wooden drums that people once used to send messages from one village to another.

The first wind instruments were hollowed-out stems of plants. When you blow down the tube or across the end of it, you get a note.

Making magnificent wooden ships like this one required thousands of acres of forest.

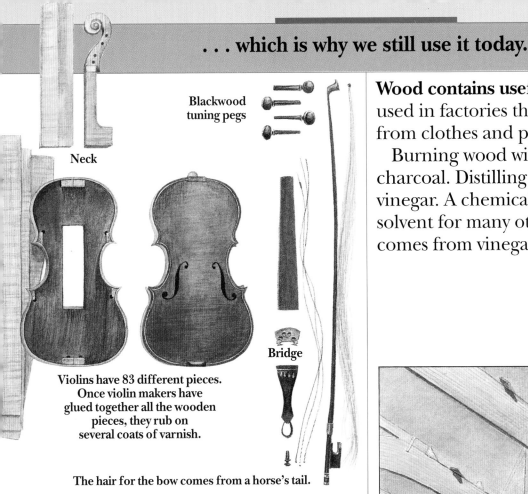

Blackwood
tuning pegs

Neck

Bridge

Violins have 83 different pieces.
Once violin makers have
glued together all the wooden
pieces, they rub on
several coats of varnish.

The hair for the bow comes from a horse's tail.

Wood contains useful chemicals. Wood is used in factories that make everything from clothes and perfumes to medicines.

Burning wood with little air makes charcoal. Distilling charcoal creates tar and vinegar. A chemical called acetone is a solvent for many other substances, and it comes from vinegar.

Plywood frame for a dome

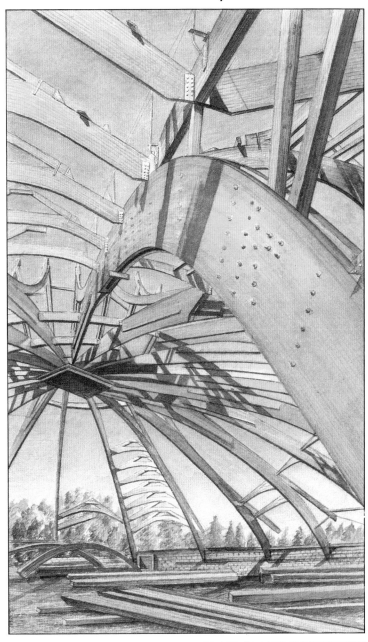

A violin maker needs to know the qualities of different woods. A violin made with one type of wood will sound different from one made with another type of wood.

Although modern buildings are made using primarily concrete, steel, and reinforced glass, wood is still a very important construction material.

New technology can make wood cheaper to use. Thin sheets of wood glued together create plywood, which is strong because the grain of one layer is at right angles to the grain of the layers above and below it. Fiberboard and particle board are made by mixing tiny shreds of wood with glue.

Today's wooden structures can be much larger than the trees from which they are made. By bending many thin strips of wood to the correct shape and gluing them together, builders make long wooden beams for bridges, sports stadiums, and other huge structures.

Trees are cut down to make pulp for paper.

If you tear a sheet a paper, you can see some of the tiny wood fibers that were pulped to make the paper. We use paper for all sorts of things, including book pages, wrapping paper, and tissues. Money, cardboard, wallpaper, even some items of artwork are made from paper. In some countries, however, schools don't have books because paper is a rare commodity.

Trees must be turned into wood chips to make pulp for paper. To keep up with the great demand for paper products, the timber industry uses highly mechanized chip mills. These mills don't require many people to operate them. One mill can turn 100 truckloads of trees per day into chips used mostly for making paper.

In some places in the United States, loggers still cut down trees with chainsaws and float them downriver to mills.

Some surprising things are made of paper.

Paper can be rough or smooth, for drawing or painting. People write letters on fine, light paper. To prevent forgery, dollar bills are made of especially strong paper that is difficult to duplicate. Paper that is as absorbent as a sponge and as soft as velvet is made into tissues and baby wipes. Paper can act as a filter for tea leaves or coffee grounds.

Baby wipes, paper towels, and tissues are all made of paper.

Cardboard is made of paper too. Pressing sheets of paper together creates cardboard, which is used for transporting fragile or heavy objects. It can be water-  proofed so that liquids such as yogurt, fruit juice, or milk do not leak out of their cartons. Paper can be reinforced to make boxes, briefcases, and bookshelves.

What did people write on before paper was invented? Early people painted pictures on the walls of caves using colors made from rocks and soil. Later, they wrote on things that would last: pottery, bones, tree bark, clay tablets, even the walls of their temples.

In this workshop in ancient Rome, scholars engrave inscriptions in stone, letter by letter.

The first books were written on animal skins.

The invention of parchment made it possible to make books. In Asia Minor, the people of Pergamon worked out the best way of drying calf, sheep, goat, and deer skins to create parchment. They cleaned, scraped, split, stretched, and dried the skins. They polished them with pumice stone until the skins were smooth sheets, thin enough to handle easily, but thick enough for people to write on both sides without the ink soaking through.

Ancient people scraped animal skins so thin that light shone through the parchment.

In the Middle Ages, books and rolls of parchment were so precious that they were kept in libraries where they were chained to the shelves. Only the richest people owned their own books. Most books belonged to monasteries, where the monks copied them by hand. The monks decorated each page with beautiful pictures, called illuminations, which made the books even more precious.

The ancient Egyptians made rolls of papyrus to write on. Egyptian scribes wrote in letters called hieroglyphs, a series of pictures rather than letters. With paintbrushes dipped in ink, they drew the hieroglyphs on long scrolls of papyrus, which is a kind of reed that grows on the banks of the Nile River. To make papyrus, the Egyptians peeled and cut the reed's stalk into fine strips and arranged the strips into layers in trays. Then they beat the layers flat and left them to dry under a heavy weight. They made the scribe's scroll by sticking several of these flat sheets together.

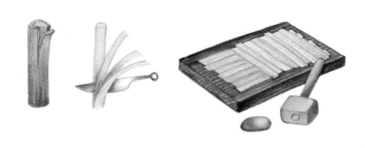

Wasps were the first papermakers.

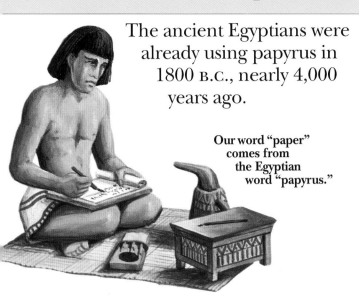

The ancient Egyptians were already using papyrus in 1800 B.C., nearly 4,000 years ago.

Our word "paper" comes from the Egyptian word "papyrus."

Much later, in A.D. 105, a Chinese man named Tsai-Lun invented paper after watching wasps build a nest.

The wasps tore off shreds of bamboo and moistened them with saliva to make a pulp, with which they made their nest. They shaped the pulp to form cells for their eggs. As the mixture dried, it hardened and kept the eggs safe. Tsai-Lun imitated the wasps. He crushed bamboo and mulberry wood in water to make a liquid pulp that he filtered through a sieve and left to dry in the sun. From this tangle of wood fibers, the first sheet of paper was made. The Chinese kept the discovery a secret for a long time.

The secret of paper was eventually passed around the world from one culture to another. In 751, during a war, the Arabs captured some Chinese papermakers and learned from them how to make paper. The Arabs changed the Chinese method by adding cotton and linen rags instead of wood to the mixture. Soon the khalifs, the kings of the Arab Empire, had libraries full of books. In the Middle Ages, crusaders saw paper during their travels in the East and brought it back to Europe.

Old rags were cut up and left in a damp cellar until they began to rot. Then they were put to soak in a vat of water. For hours, huge wooden beaters broke them into tiny pieces. These bits mixed with the water to make a runny paper pulp.

A river's running water turned the mill wheel. As it turned, the wheel worked the heavy beaters that made the pulp.

The creation of a sheet of paper

Workers plunged a mold (a fine sieve) into a vat of paper pulp. Then they shook it to spread the pulp evenly. They put this layer of pulp between two pieces of felt and pressed the layers in a huge press until all the water was squeezed out.

By the middle of the first century, the Chinese had been printing paper, but the technology took until the Middle Ages to reach Europe. About 1450, in Germany, Johannes Gutenberg invented a way of printing by using letters on small metal bars and a special printing press. The press could make many copies of a book. People wanted more books, but the papermakers didn't have enough rags to make the amount of paper they needed. In the 18th century the Chinese method of making paper was rediscovered, and today most paper is made from wood pulp. The best quality papers, called rag stock, are still made from rags.

In a factory, machines do the work that the paper mills used to do, but instead of making paper one sheet at a time, modern machines turn out a half mile (800 m) of paper every minute—nearly eight times the length of a football field.

How can we plant enough trees to keep up with the paper demand? Saplings are planted to replace the trees cut down. But forest trees take a long time to grow to maturity—up to 60 years for a beech, for example. Most tree farms are planted with pine because pine trees can be turned into pulp after 20 years.

Scientists are experimenting with making paper out of plants such as hemp, flax, and esparto grass. To avoid waste, some papermakers recycle paper out of discarded newspaper and other paper products, but the process is expensive. Recycling paper saves trees, reduces air pollution, and keeps more paper wastes from ending up in landfills.

Pulp is poured onto a wire mesh, which moves like a conveyor belt (1). Rollers squeeze the pulp (2), and then it moves between heated cylinders (3). The paper is carried along on a bed of felt.

As the paper passes through the press, heated rollers dry it, smooth it, and roll it onto a bobbin (4).

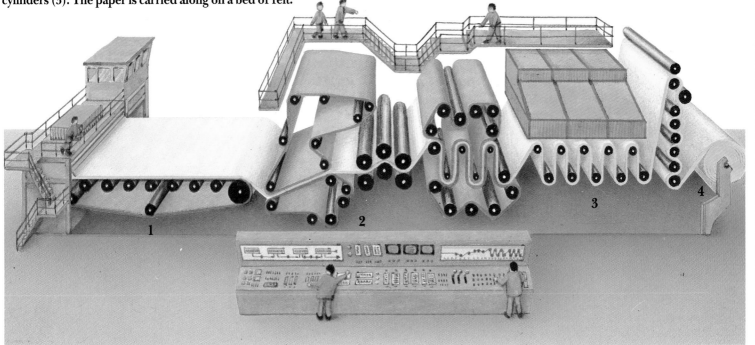

As soon as they learned to hunt, people began to use animal skins to make leather. Leather is strong and waterproof but not rigid. Because it is skin, it breathes. As  time passed, people used leather to make clothing, shoes, book covers, and horse saddles. Today leather is becoming a luxury because inexpensive plastics have replaced leather as a construction material.

Craftspeople who worked with leather were specialists. Some tanned the skins of cows and oxen, while others worked with the softer skins of sheep and goats.

Almost any kind of animal skin can be made into leather. Most of the leather you see today is made from the skins of pigs, calves, sheep, or goats. Because crocodiles and sharks are rare animals, leather products made from their skin are expensive. In many countries it is illegal to own materials made from endangered or threatened animals such as these. Native people such as Inuits of Alaska and Greenland once made their clothing from bear and seal skins by stretching it on a frame to dry. People in Tibet make their boots from yak skin.

Before it can be tanned, skin has to be scraped clean to get rid of any flesh, fat, and hairs.

. . . to make leather that is dyed and crafted.

Before leather processing factories, people trampled skins in great vats to soften them.

How is a skin made into a piece of leather?

Animal skin contains water; it would rot if the water was left in it. To make skins last, tanners put the skin through a series of processes. First, they soak a skin in water to soften it. Then they put the skin in lime to loosen the roots of the hairs. Next, the tanners remove the hairs and scrape both sides of the skin clean. Finally, they tan the skin by soaking it in a vat of tannin, a chemical found in the bark of trees such as oaks and chestnuts. Tanners also use alum, formalin, and chrome—all these substances prevent the skins from rotting.

In the Middle Ages the finest leather came from Spain, especially from Cordoba. The craftsmen who worked with Cordoban leather were known as 'cordwainers'.

Once the leather has been tanned, the artisans grease it to make it waterproof, and perhaps dye it different colors.

Leather workers

Cobblers made shoes and mended them when they wore out. Saddlers still work by hand, making saddles, reins, harnesses, and other leather parts of a horse's tack. Glovers make leather gloves, and bookbinders put beautiful leather covers on books. Applying leather to desks and tables is yet another craft.

All these types of work need skill, and things made by hand are always more expensive. Today, most leather bags, shoes, clothes, and furniture is made on a production line in a factory.

Feet sweat less in leather boots.

Leather bags last a long time.

A leather saddle is soft and comfortable.

Leather horse collar

Leather purse

Spanish craftspeople learned from the Arabs how to make beautiful gold-tooled leather book covers.

The first clothes were animal skins and furs.

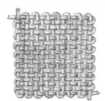

Animals have fur, feathers, or scales to protect them from the heat of the sun and from the cold, wind, and rain. People have hair on their head, but otherwise their bodies are bare and unprotected. But human beings have clever minds, nimble fingers, and the ability to make things.

If you had been born 100,000 years ago, your parents would have wrapped you in animal skins to protect you from the harsh weather, prickly plants, and rough ground. If you had been born thousands of years later, your parents would most likely have used clothing made from woolly animals like sheep.

To have enough meat to eat, skins to keep warm, and bones to make tools, people began to keep herds of animals near their homes. These animals became used to living near people. Tufts of soft wool from sheep would catch on brushes and people collected it. They rolled the wool between their fingers and pulled it out into long strands. They had invented spinning.

Does wool come only from sheep? Many animals all over the world have woolen coats, though they are not so easily farmed as domestic sheep.

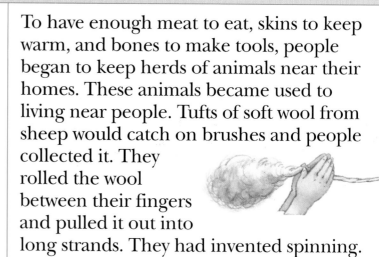

Camels and dromedaries in Asia and Africa, yaks in Tibet, and llamas in South America are mostly used to carry things, but they are also kept for their coats, which are a valuable source of wool.

People began to keep animals for their wool.

The wool from a cashmere goat is soft and warm.

The fur from angora rabbits is sometimes mixed with wool.

Two important inventions were the spindle and the loom. Civilizations all over the world used the spindle, which looks like a long toy top. Spinners attached a strand of wool to the spindle and set it spinning. As the spindle hung down, it wound the wool into a thread.

A long, fine strand of yarn formed around the spindle. The invention of the loom meant that weavers could create longer and wider pieces of cloth more quickly. They stretched threads along the loom, from one end to the other.

Ancient spindle and woven clothing from northern Europe

A loom used in Egypt about 4,000 years ago

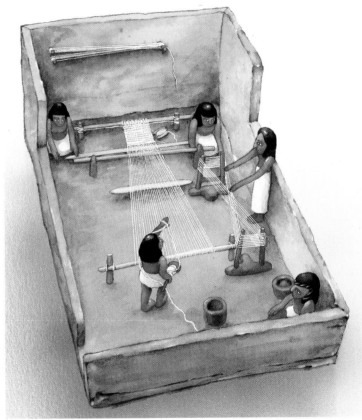

Llamas, alpacas, vicuñas, and guanacos supply wool to the Indians who live in the mountains of South America.

Then they threaded the yarn across the loomed thread, over and under, over and under. They used a shuttle to carry the thread from right to left, and then from left to right. The weavers could make complicated patterns by using different colored threads.

Shepherds and their dogs look after sheep.

Shepherding is skilled work both for shepherds and sheepdogs. In the past, shepherds and their dogs spent the whole year living with the flock. The shepherds

taught their dog a range of whistles, gestures, and words to tell it to lie down, run, or go left or right. The shepherds relied on their dog to help herd the sheep and guard them. In some countries, the dogs even had to fight off predators.

Sheep usually live in a field during the warmer months. They may spend the winter months in pens.

Healthy sheep have pure, thick fleeces.

Sheep used to be moved to fresh pastures for the summer. In some countries, the fields are too dry to feed the sheep in summer, so the shepherds took the flocks up into the mountains where the grass is fresh and sweet.

Shepherds dressed the rams in bells and bright pompoms so that the sheep didn't get lost.

Many kinds of sheep

Farmers are always trying to improve their sheep herd. Some animals are kept for their milk, which is made into cheese. Others provide meat. But many sheep are kept for their wool. The best wool comes from healthy sheep with fine, thick fleeces, like the merinos. Huge flocks of merinos roam the countryside in Australia, New Zealand, and South America. In New Zealand there are 13 times as many sheep as there are people!

MERINO

What makes sheep's wool so special? Sheep's wool is bouncy and soft—every hair is curly and covered with tiny scales that protect it. A wool sweater

HEBRIDEAN

must be washed carefully so that the scales are not damaged—otherwise the sweater will shrink. On each side of the hair are glands that produce the oil and sweat that make the wool waterproof. This oil produces lanolin, which is made into hand lotion and hair gels.

An Australian sheep farmer examines the fleece of one of his sheep. If he finds any parasites—small creatures that live on the hairs—he will dip the sheep in a special bath to get rid of them.

Before you can wear a woolen hat or sweater, the fleece must be processed into wool and then into cloth.

Sheep are shorn in spring. Shearing is done by hand, with scissors or shears, or with an electric shearer.

Because different qualities of wool exist in each fleece, the wool must be **graded** and separated by hand.

The wool is washed several times to remove any fat, sweat, or bits of grass.

To untangle the hairs, the wool is **carded** with special combs. Now it is smooth and ready for spinning.

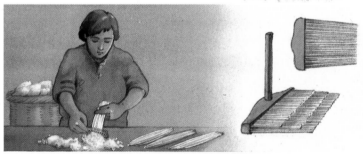

Combing the wool makes it even smoother and fluffier. This wool can be spun into long fine threads or can be used to make a soft filling for mattresses and cushions.

The **spinning wheel** was invented to speed up the process of making threads. Modern machines spin thread finer and faster.

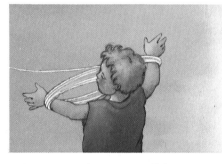

The yarn is wound into skeins, and then into balls or onto bobbins. Today all this is done by machine.

Wool can be dipped in dyes made from plants or chemicals to produce every color of the rainbow.

In the past, all knitting was done by hand, and workers used to knit night and day.

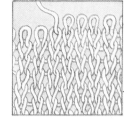

In the 16th century, a young Englishman, William Lee,

decided his girlfriend should not have to work so hard, and he invented the first knitting machine.

You only need two needles and a ball of wool to knit yourself a scarf!

For complicated knitting patterns, cards with holes punched in them guide the machine.

Where does silk come from?

Silk fabric is soft, shiny, and luxurious. It's hard to believe that a small caterpillar called *bombyx mori* creates the silk threads. The caterpillar spins silk to make into the cocoon that it will live in while it turns into a moth. More than 4,000 years ago, the Chinese learned how to keep these silkworms. They used the harvested threads to weave wonderful fabric to make clothes, flags, and banners.

The *bombyx mori* eats only mulberry leaves. Later it turns into a large white moth. In early summer the female moth lays about 500 eggs, each about the size of a pinhead. The silk farmer keeps the eggs cool so that they do not hatch too soon. They are allowed to hatch the following spring, just when the tender new mulberry leaves appear. For the first four weeks of its life, the caterpillar eats as many delicious young leaves as it can, gradually growing fatter and fatter. After a month, it is very plump—8,000 times the length it was when it hatched.

Then it stops eating and spins its cocoon. A sticky thread, which may be more than a half mile (1 km) long, oozes from the silk glands on its head. Then, inside the cocoon, the caterpillar goes to sleep.

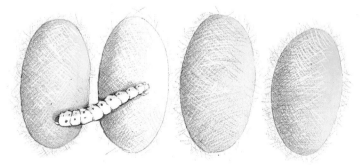

China produces and exports more silk than any other country in the world. Bales of silk are flown to Europe, the United States, and other countries in a matter of hours after the silk is harvested. Japan and India also produce silk, and in these countries silk is used to make kimonos, saris, and other fine clothes.

Workers transport cocoons by water canals to the factories (left). An orchard of mulberry trees in China (below). Farmers have learned how to grow dwarf mulberry trees. The trees are short and workers can pick the leaves easily.

After two weeks inside the cocoon, the chrysalis has turned into a moth and is ready to emerge. It uses its saliva to soften the silken walls, and then forces its way out.

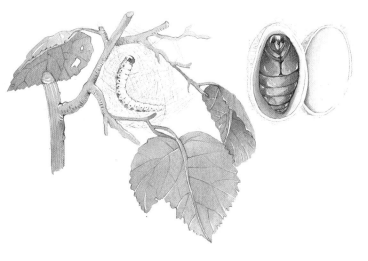

At first the moth is damp and crumpled, but after a few moments in the open air, it is ready to stretch its wings. Its life is short—it has just enough time to lay its eggs—and the life cycle of the silkworm begins again. But on a silk farm, workers allow only a few moths to emerge and lay eggs. They set aside most of the cocoons for the manufacture of silk.

Unraveled cocoons produce silk thread.

Unwinding cocoons

When harvesting silk, farmers need to make sure that the moths don't try to get out of their cocoons.

Emerging moths tear the precious thread and leave behind a heap of tiny silk fragments. To quickly kill the chrysalis inside, farmers place the cocoons into a hot-air bath. Then they place the cocoons in boiling water to soften them. At the same time, they pass a brush over the cocoons to loosen the end of the thread. Workers then put the cocoons on a swift, which is an unreeling apparatus that unravels the cocoons just like a ball of wool. The workers feed the thread through pulleys so that it winds evenly around the swift without getting tangled. At the end, a skein of cream-colored raw silk is ready to be taken off the swift.

A swift unravels the thread from the silkworms' cocoons.

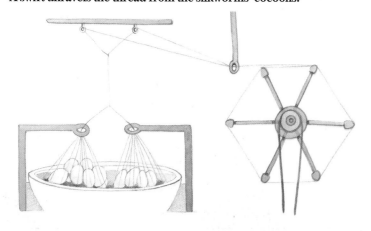

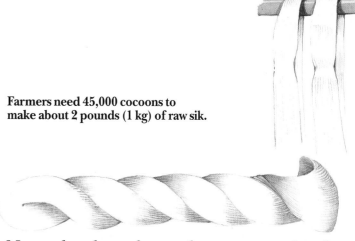

Farmers need 45,000 cocoons to make about 2 pounds (1 kg) of raw silk.

Next, the threads are thrown, or twisted, to keep them from becoming tangled. Now they are ready for weaving. Woven silk can be colored by soaking it in dye, or a pattern can be printed on it.

Silk is a symbol of elegance and luxury. For a long time, only rich nobles and their families could afford silk. They dressed from head to toe in it—even their shoes and stockings were made of silk. Some fabrics were embroidered with gold and silver thread. Castle walls were hung with silk tapestries.

How did silk come to the West? Travelers from the Roman Empire were amazed by the silken cloth that they saw in China. Soon people all over Europe wanted to buy silk. But there was a long and perilous journey along the famous Silk Road before the bales of silk could reach the West.

Traders brought silk from China to Europe.

Several months of traveling lay ahead for the camel trains, laden with silk, spices, and porcelain. The journey took the caravan over mountains and across deserts, where bands of robbers prowled. Once the merchants reached the Mediterranean Sea and crossed it safely, the silk was delivered to the market. People bought it and made splendid clothes.

The Silk Road was a caravan trail that stretched from the middle of China across Central Asia to the shores of the Mediterranean Sea.

Silk weaving became important in Great Britain in the 16th century, when many weavers arrived from France, Belgium, and Holland. Some weavers settled in London, around Spitalfields. They kept their looms in their houses, which made the neighborhood very noisy. The whole family, including the children, helped with the work.

People all over the world wear clothes made of cotton. Light cotton keeps you cool when the weather is hot, and padded cotton keeps you warm when it's cold. Cotton can be dyed any color. Unlike wool or silk, a plant produces cotton. As the cotton plant grows, its fruit becomes white and fluffy until ultimately field workers pick it and make it into cloth.

Cotton seeds are sown in the rainy season. Two months later, the plants sprout thick branches covered with beautiful flowers. As the flowers fade, small green fruits replace them and eventually turn into the seed heads that contain the cotton.

Picking cotton in Africa (below)
Planting cotton: In certain areas of Africa, the entire village helps (right).

Some of the world's cotton comes from Africa, specifically from Egypt, Sudan, Zimbabwe, and Mali. In certain areas of Africa, everyone in the village helps with the harvest. The villagers gather ripe seed heads by hand and fill baskets.

In the marketplace, the villagers tip the cotton out of the baskets and pile it up in white heaps. They keep part of the crop so that they can sow the seed during the next rainy season. They press some of the seeds to make into oil. Most of the harvest is sold to countries with cotton mills, the factories where cotton is spun into thread. Before the cotton can be sent overseas, it has to be compressed into bales—otherwise it would be too bulky to transport.

In some African countries, gathering cotton (left) and picking cotton (below) are still done by hand.

For thousands of years, people have grown cotton for cloth.

Cotton will grow wherever the climate is warm and damp. For more than 5,000 years, people in Asia and Central America have grown cotton to make cloth. In the United States, cotton clothes were still a luxury in the 16th century.

In 1793 Eli Whitney invented the cotton gin, which removed the seeds from the cotton fibers. This made cotton production cheaper and faster, but created the need for more workers to keep up with the demand for cloth and other cotton products. To keep up, plantation owners in the Southern U.S. kept hundreds of slaves—people taken from their homes in Africa and made to work in the cotton fields. The owners forced the slaves to work nonstop from sunrise to sunset. By 1860 four million slaves lived in the United States.

Cotton continues its popularity. Today more than 70 countries grow cotton. The largest producers are China, the United States, and some of the countries of the former Soviet Union. These three areas

provide half the world's cotton. India is the next largest producer.

Cotton cannot be grown in Europe or Japan, so these countries import it from the cotton-producing countries and spin it in their mills.

Plantation life: slaves were forced to pick cotton (below). The abolition of slavery was one principle over which the American Civil War was fought. Today, machines pick cotton (above).

From cotton plant to cotton cloth

Several times through-out the growing season, special airplanes spray the cotton plants with chemicals to get rid of  caterpillars, flies, and weevils that eat the leaves and seed heads. At harvest time, machines that act like giant vacuum cleaners suck up the heads. Leaf pieces are sometimes gathered too, but are sorted out later. One machine can do the work of a hundred people.

Cotton fibers are separated from the seeds.

In the cotton mill, the cotton is dried, and the seeds are taken out. At this stage each cotton bale weighs 477 pounds (217 kg).

The cotton is still a mass of tiny threads, which a machine disentangles. Another machine combs the cotton, pulls it out into a soft hank, and finally spins it. The more tightly the machine spins the thread, the stronger it is. Cloth made from this cotton thread is not really white. The thread has to be cleaned and bleached. It can then be treated so that it does not crumple too easily or shrink the first time it is washed. It can also be made waterproof.

The cotton industry began in the 18th century, when the first spinning machines were invented.

Today cotton is sometimes mixed with other types of fiber, usually either wool or synthetic fiber such as nylon, polyester, or rayon. **Synthetic fibers** are artificial, not natural. They are made from wood pulp, coal, or oil. Mixtures of cotton and other fibers shrink less than pure cotton and tend to last longer.

The bales are cleaned and mixed together.

The cotton is combed and turns into a long, soft ribbon.

Several ribbons are put together to make a hank.

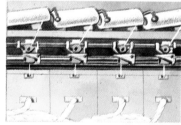

The hank is pulled and twisted tightly to make a thread.

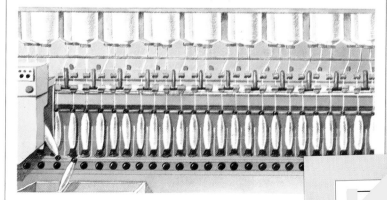

The thread is wound onto bobbins or spindles.

Modern factories use chemical dyes instead of natural colorings. People used to use plants to make beautiful colors: blue from indigo, red from madder, yellow from weld.

To print large pieces of material, producers may use screens like stencils to put the dye onto the cloth. When the dye is poured onto the screen, it squeezes through the unblocked areas, coloring the cloth below it.

Printing roller

People in China wear winter clothes made of padded cotton.

How is the pattern put on cloth?
Producers can paint cotton material with brushes or print on it with rollers. They can also print patterns on it with stamps.

Not all cotton material is the same. Poplin is smooth; percale is glossy; muslin is lightweight; toweling is fluffy; satin is shiny; and velvet is soft. People knit or crochet with cotton yarn. Some use it to make lace or to embroider beautiful patterns on linen and canvas.

One of the most common kinds of cotton was originally only grown in South America. Peruvians wove brightly colored cotton ponchos and rugs on simple looms.

Cotton is grown in countries around the world.

Linen is made from the fibrous stalks of flax plants. It is the oldest known fabric. Archaeologists have found pieces of prehistoric linen that have not rotted.

Agave grows well in hot countries like Mexico. Its thick leaves are used to make a fiber called sisal. Sisal can be twisted to make string, ropes, rough carpets, and hammocks.

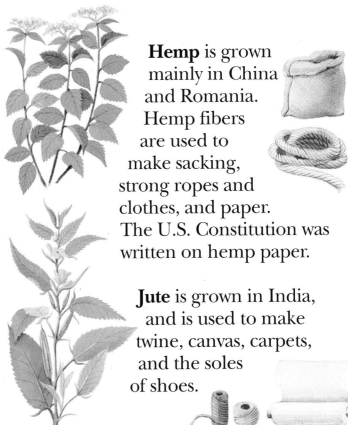

Hemp is grown mainly in China and Romania. Hemp fibers are used to make sacking, strong ropes and clothes, and paper. The U.S. Constitution was written on hemp paper.

Jute is grown in India, and is used to make twine, canvas, carpets, and the soles of shoes.

Oil formed over millions of years . . .

We don't live in the Stone Age or the Iron Age—sometimes it seems as if we live in the **Plastic Age.** Made from oil, plastic has become one of our most common materials. Oil, a hydrocarbon like coal, began to form at the bottom of the sea long before dinosaurs walked the Earth. For millions of years, tiny little algae and microscopic animals called plankton drifted down to the seabed when they died. As they sank to the bottom, their bodies slowly rotted. First they turned into a kind of sludge. Over millions of years this sludge turned into the thick black liquid we call oil.

People have known about oil for thousands of years. They first used it to make things waterproof, but now it is the raw material for many industries, and at present we all depend on it for fuel.

Noah's Ark

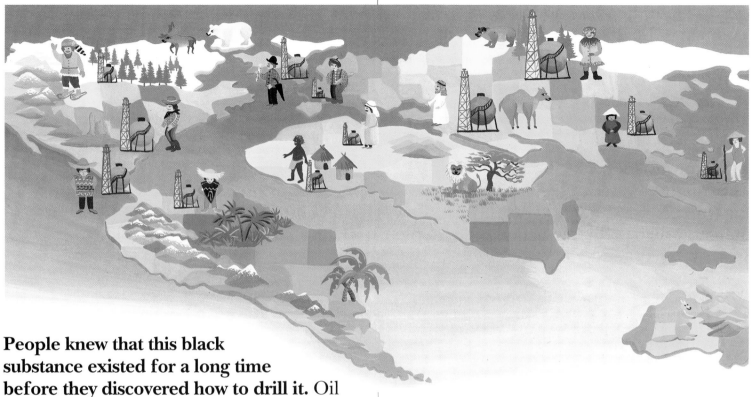

People knew that this black substance existed for a long time before they discovered how to drill it. Oil is even mentioned in the Bible: Noah lined the inside of the Ark with pitch to make it waterproof. Pitch, a type of oil, rises to the Earth's surface in black lakes. About 1850, oil was in great demand to use as fuel for lamps. American pioneers found oil in the West, quite by chance, when they were digging in the ground looking for salt. In 1859 Colonel Drake sank the first oil well in Titusville, Pennsylvania. Soon the well was producing more than three tons of oil each day.

The rush for "black gold" began in the United States. Because of its color and worth, people called oil "black gold" and raced to dig their own wells. Later on, people began to look for oil in other parts of the world. Today it has been found under the sea, as well as underground in some countries where sea covered the land in prehistoric times.

Today oil companies employ geologists and other scientists to search for oil. These companies not only drill and extract the oil, but also refine and sell it. They transport the oil and gas by tanker or pipeline from the oil field to the places where it will be used.

An early oil well. The wooden tower, called a derrick, was built over the well. Inside the tower was a pulley and a heavy weight that was used to crush the rock.

Much of the world's oil is buried beneath the seabed.

To drill into the rocks under the sea, oil companies build huge platforms made of concrete and steel. These rigs are built in shipyards, and then tugboats tow them out to sea. Oil rigs are tall—but the part underwater is even larger than the tower rising above the sea. Concrete holds the rig firmly on the sea bottom, so that even the roughest waves can't knock it over. Oil rig workers use a tough steel drill to penetrate the hard rock of the seabed.

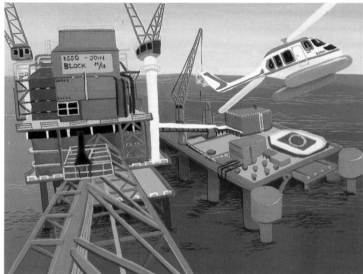

Sometimes the rock is too hard even for the steel drill. Then workers use a drill edged with diamonds. When they've bored a hole in the rock, the workers drop a pipe into the hole; oil comes up the pipe under pressure. Cameras watch over the underwater equipment. If anything needs to be repaired, divers go down and mend it.

About 100 people are needed to operate an oil rig. The work is hard, so teams change every two weeks or so.

A new team arrives by helicopter.

Workers use walkie-talkies.

The control room

Changing a drill head

Crude oil goes to the refinery to be treated.

Crude oil—straight from the earth—can't be used until it has been refined. Oil used to be carried to the refineries in barrels, which is why oil is still measured in barrels. Today huge oil tankers transport the crude oil by sea, or massive steel pipelines, like the Alaskan Pipeline, transport it over land. At the refinery, oil is turned into various fuels such as kerosene for aircraft, gasoline for cars, diesel for buses, and heating oil.

Naphtha is a sort of gasoline that has been treated in a chemical factory. It is used for making many things, including detergents, medicines, and plastics. As the world's population increases, the demand for oil goes up. Scientists believe that the Earth's oil will be used up before the middle of the 21st century. This has led some scientists and engineers to explore **alternative sources of energy,** such as natural gas, solar, wind, and water power.

Chemists study atoms through powerful microscopes that magnify things millions of times their actual size.

The magic of chemistry

Take a look at all the plastic items around you, from the storage containers in the kitchen to the inside of a car. How are these different kinds of plastic made? Everything in the universe is made up of tiny particles of matter called atoms. Like interlocking building blocks, atoms lock

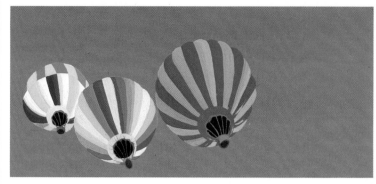

together in different ways to make different objects. Chemists make plastics from the gases that come from oil when it is heated.

Chemists have discovered how to take oil atoms apart and lock them back together again in new orders—making thousands of different chains of atoms. Each of these chains forms a different kind of plastic.

Plastic can be made to look and react like almost anything: wood, glass, wool, metal, or rubber. It can be transparent, but it is less breakable than glass. It can be flexible or rigid. It is lighter than metal. Plastic can keep out heat or cold. It can be used to make soft, warm material or water-proof fabric that keeps you dry.

Today, surgeons use plastic to replace all sorts of parts of the human body, including joints, arteries, and heart valves.

Plastic is an everlasting material.

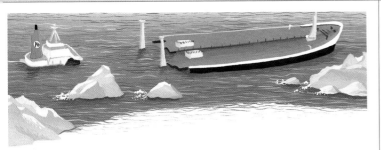

An oil tanker has been wrecked near the coast. The oil that is spilling into the sea pollutes the beaches and kills thousands of birds, fish, and sea mammals.

What happens to the tons of plastic that we throw out every day? They stay here! Natural things eventually break down and feed the soil. But plastic takes hundreds of years to begin to break down. For example, disposable diapers, which are one-third plastic, need 200 to 500 years to break down in landfills.

Plastic is not as easy to recycle as glass. Recently engineers have figured out how to melt down plastic bottles and shred other plastics in order to make things such as videotapes, toys, seat belts, and clothes. Recycling plastic items keeps them from clogging landfills.

Plastic is part of our future. Plastic manufacturing is important to space exploration. Some plastics are 10 times as hard as steel and can stand up to high temperatures—rocket heads and fuel tanks are made of plastic.

Sea animals can get tangled up in or swallow trash thrown into the sea—which can endanger their lives. Trash dumped at sea may wash back up onto beaches.

Intriguing facts, activities, games, a quiz, a glossary, and addresses of places to visit, followed by the index

Whitewash is made from lime, which is made by heating limestone. People whitewash their walls and fences to protect them from damp weather.

The chalk you use to write on the blackboard is made from gypsum. This soft mineral is ground to powder and pressed into sticks. Chalk can be made various colors.

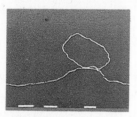

Talc, the softest mineral of all, makes the talcum powder that some people sprinkle on themselves after a bath or shower.

Prehistoric people chipped **flint** until it had sharp edges. They created simple tools and weapons such as knives, axes, scrapers, daggers, and arrowheads.

Gold is a precious metal. In the Middle Ages alchemists, who were part scientists and part sorcerers, tried to turn ordinary metals into gold—without success. We now know there is gold in sea water (more than 900 million tons of it), but it would cost more money to extract it than the value of the gold. The largest gold nugget, called Welcome Stranger, was discovered in 1869 in Australia. It weighed 156 pounds (70.8 kg).

The legend of the imaginary country of Eldorado

The word Eldorado comes from the Spanish for golden man. The conquistadores who traveled to South America heard of a local ceremony in which a man was covered with gold dust. Such tales led them to believe that the Amerindians had enormous gold stores hidden in the forest. The Spanish searched everywhere for this treasure, but never found it.

Roman women wore makeup made from natural substances. They had a white paste made from powdered chalk, which they spread over their face and arms. They used red ocher, which is a kind of clay, on their lips and cheeks, and they made mascara from soot.

A carborundum stone is a hard, slightly rough stone. A person can keep a good edge on the blade of a knife or an axe by grinding it regularly on the stone.

Pumice, solidified lava foam, is used to stonewash jeans, as an abrasive in soaps, and for concrete building blocks.

Magnets are made of iron. The magnetized needle in a compass always points toward the north.

Mercury, also called quicksilver, is bright and shiny. It is the only metal that is liquid at ordinary temperatures, so thermometers and barometers often contain mercury. When mercury is heated, it turns into a poisonous gas. It is also used to purify other metals such as silver and gold, and to make dyes. The ancient Greeks mixed it with other metals such as gold to form alloys known as amalgams.

Lead, a heavy metal, is the metal used by plumbers (*plumbum* is Latin for lead). Mixed with tin it makes good solder. Fishermen used to tie lead weights onto their fishing line, making a "plumb line," until they realized that the lead was poisoning swans that accidentally ate the weights. Hunters and soldiers used to load their guns with lead ammunition.

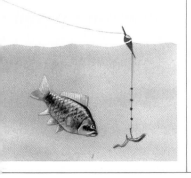

Space factories of the future

It took many millions of years for Earth's resources to form. Humans are using these precious resources too quickly. Scientists are trying to solve this problem by developing alternative energy sources.

In space, objects are weightless. They float around in zero gravity. In the future there will be orbiting laboratories where special, lightweight alloys can be made because there is no gravity.

It will also be possible to blend together different substances, like steel and glass, to create materials with exciting new properties.

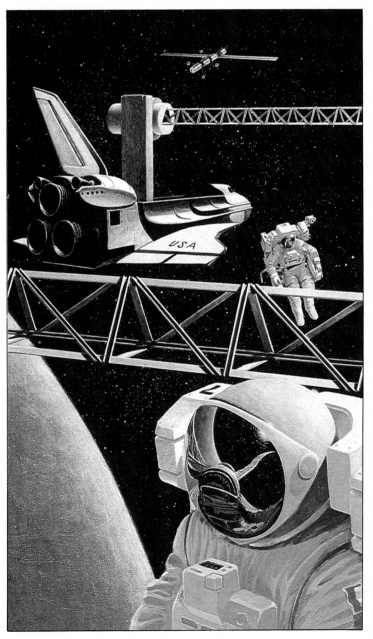

Aluminum, a good conductor of electricity, is used to make electric cables. It is also used to make saucepans because it conducts heat well.

Titanium is a light metal used to build aircraft and racing bicycles.

Silk, mixed with other fibers, is used to make display parachutes and the tires used by racing bicyclists.

Plastic Lego building blocks first appeared in 1955. They were invented by a Danish carpenter who was interested in designing toys. The name Lego comes from the Danish words *leg godt,* meaning play well. The largest collection of Legos in the world is at Legoland in Denmark. The display is full of enormous constructions all made with interlocking Lego building blocks.

In 1900 in Great Britain, a man wanted to give his sons a toy with which they could build things. He thought up the idea of metal strips full of holes that kids could join together with nuts and bolts. Then he added pulleys and other parts to his set so his sons could build a miniature crane—he had invented **the erector set.**

Dolls used to be made of all sorts of different materials, including wood, papier-mâché, terra-cotta, wax, ivory, and china.

■ What happens in a plastics factory?

How are Lego building blocks made? Plastic granules are poured into an injection mold. As the granules move along, they are heated until they melt. The hot liquid is injected into the end of the mold. After it cools, the mold is opened, and a building block emerges.

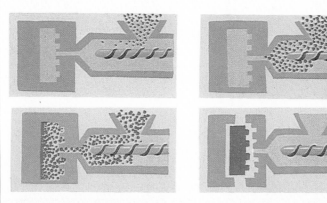

What about plastic bowls? Objects like bowls, buckets, plates, and mugs are made by **heat molding.** Plastic powder is tipped into a mold that is closed and heated. When the mold is opened again, the bowl has been shaped—like a waffle in a waffle iron. An ejector pushes it out from below.

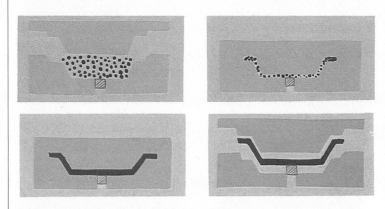

And hollow things like bottles? They are blown. A tube of hot plastic is put into a mold. The mold is shut, and compressed air is blown into the tube. The tube swells up like a balloon and presses against the sides of the mold. The mold is opened, and the new bottle just needs to cool before it is ready for use.

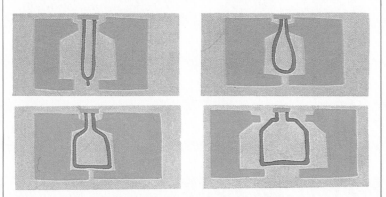

The first **imitation diamonds** were invented about 1745 by a jeweler in Paris, called Strass. He found out how to make a kind of glass containing lead and metal oxides, which sparkled just like real diamonds.

In 1748 Fraçois Fresneau, a Frenchman, invented **waterproof boots.** He coated some boots with latex, a kind of sap. In 1823 a Mr. Macintosh, in Great Britain, discovered a way to make cotton material waterproof by using a mixture of rubber and turpentine.

The idea for enormous buildings supported on a framework of metal girders—**skyscrapers**—was inspired by accident. Mrs. Jenny, the wife of an American architect, dropped a heavy book onto a birdcage, without harming the cage. Her husband realized that the rather fragile-looking object was really very strong, and that tall buildings could be designed in the same way.

■ Where do colors come from?

Before modern chemical dyes, people used plants, minerals, and even animals to make dyes.

Black comes from carbon, or oil smoke (lampblack).

Red is made from cochineal, a crimson dye obtained from the powdered eggs of an insect that lives on a Mexican cactus. Rusty iron gives another red. Minium, the red color used in medieval manuscripts, comes from lead. Anatto, from the fruit of the bixa tree, gives a strong orange color, and safflower makes a bright red.

Brown can be made from any soil that contains either iron oxide or carbon. Colors made like this are called ocher. Onion skins, chicory roots, tea, and walnut skins all produce various shades of brown.

These four flowers can be used to make dyes. Can you find their names? They are hidden in the text in the left and right columns.

From left to right: safflower, bixa, weld, indigo.

How are designs printed on material? Sometimes designs are printed with rollers on which the pattern is embossed. Wax protects some of the material while the rest is dyed. Some machines can print 20 different colors, one after another. To print large pieces of cloth, the dye is put on through screens, which work like stencils, keeping the dye off of some parts of the design. To screen print a multicolored design, you use one screen for each color.

Cloth can be painted by hand, using brushes and stamps of different shapes and sizes. Or it can be tied in various places and dipped in dye, like the blue cloth below.

A large brush and a stamp

Yellow is made from many plants—broom, heather, bracken, weld, nettles, and alfalfa. But it usually comes from the roots of turmeric, an Indian plant.

Green is produced from the verdigris (oxide) that forms on copper when it is exposed to air—the color appears on copper roofs and statues. It is bright green and poisonous.

Many **blue** dyes are made from minerals, especially copper or coal. But indigo, a dark blue, comes from the leaves of the indigo bush, which grows in countries with very sunny, hot climates.

Purple can be made from black currants, grapes, and blackberries.

White comes from chalk in the ground, or powdered shells—from oysters or even from ordinary chicken eggs.

Make yourself a rabbit or a kitten.

You will need a pair of old wool gloves,

some filling or cotton,

ribbon or yarn, a needle, thread, a pair of scissors, glue, felt, and buttons.

To make the rabbit:
1. Tuck in the thumb and the two middle fingers of the glove.
2. Sew up the fingers you tucked in, and stuff the glove with cotton or other filling.
3. Sew up the bottom.
4. Tie a length of ribbon or yarn around the neck.
5 & 6. Glue on pieces of felt for eyes and sew on a button for a nose.

To make the kitten:
1. Tuck in the thumb and all the fingers except the first (for the tail). Sew up the fingers you tucked in. Stuff the glove and sew up the bottom.
2. Tie a ribbon or yarn around the neck, and a shorter piece around each corner to make ears.
3. Glue on pieces of felt for eyes and sew on a button for a nose.

A game of marbles, anyone?

Marbles is an old game. Children in ancient Greece played it with acorns or olives. Four thousand years ago, the ancient Egyptians had glass marbles, but theirs were not really round. The first round marbles were made in the 18th century.

Here's how to play: Draw a huge circle on the ground and a smaller one inside it.

The smaller circle is the sun. Players stand at the edge of the big circle. Each player tries to roll a marble into the sun circle. If you miss, you lose a turn.

If you succeed, take a step forward. The first person to reach the sun circle is the winner.

■ Cross-stitch embroidery

If you like sewing, copy this alphabet to make a sampler. First sew one-half of the X's in a row. Then double back and complete the X's.

■ Papier-mâché shapes

You'll need newspapers, a large bowl, water, measuring cups and spoons, a packet of wallpaper paste or wheat paste, cookie cutters or lids, vegetable oil, paintbrushes, craft paint, and varnish.

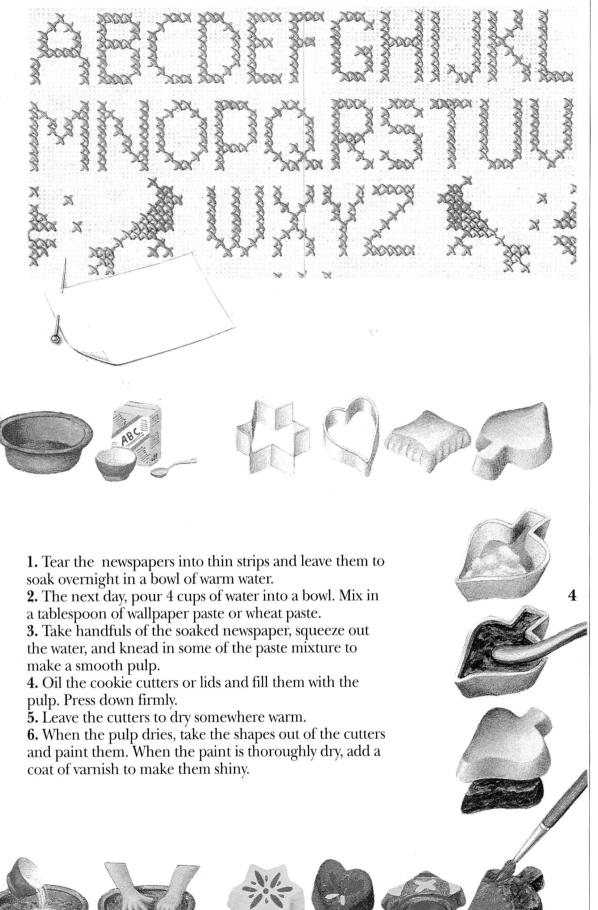

1. Tear the newspapers into thin strips and leave them to soak overnight in a bowl of warm water.
2. The next day, pour 4 cups of water into a bowl. Mix in a tablespoon of wallpaper paste or wheat paste.
3. Take handfuls of the soaked newspaper, squeeze out the water, and knead in some of the paste mixture to make a smooth pulp.
4. Oil the cookie cutters or lids and fill them with the pulp. Press down firmly.
5. Leave the cutters to dry somewhere warm.
6. When the pulp dries, take the shapes out of the cutters and paint them. When the paint is thoroughly dry, add a coat of varnish to make them shiny.

■ Quiz

Can you answer these questions? The correct answers are at the bottom of the page.

1. Metamorphic rocks are formed
a. from the bones of dead animals.
b. during violent shifts of the Earth's crust.
c. under the surface of the Earth.

2. Sapphires are usually
a. blue.
b. green.
c. yellow.

3. Bronze is
a. a precious metal.
b. an amalgam.
c. copper and tin combined together.

4. Metallurgy is
a. the extraction of metals.
b. the study of metals.
c. jewelry making.

5. The crystal used in chandeliers is
a. a type of transparent rock or stone.
b. glass with lead added.
c. a precious stone.

6. Wood is used to build boats because
a. it floats.
b. it is strong.
c. it is easy to carve.

7. Cardboard is made from
a. chopped up wheat straw.
b. wood pulp.
c. powdered animal bones.

8. Animal skins are tanned
a. to clean them.
b. to dry them.
c. to make them last.

9. The wool on a sheep is shorn
a. in the spring.
b. all year round.
c. in winter.

10. Silk is made from
a. a sticky thread produced by caterpillars.
b. the fibers of a silk plant.
c. the threads of a spider's web.

11. Cotton comes from
a. sheep.
b. the bark of a tree.
c. a plant.

12. Oil comes from
a. algae and sea creatures that have rotted.
b. decayed tree trunks.
c. magma at the center of the Earth.

13. Diesel is
a. fuel used by buses.
b. the type of oil used in oil lamps.
c. oil used in an oil-fired central heating boiler.

14. Plastic is made from
a. diesel.
b. fuel oil.
c. naphtha.

15. A geologist
a. studies rocks.
b. digs out the tunnels and galleries in a mine.
c. studies animal bones.

Answers: 1 b, 2 a, 3 c, 4 b, 5 b, 6 a, 7 b, 8 c, 9 a, 10 a, 11 c, 12 a, 13 a, 14 c, 15 a.

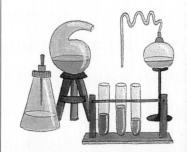

■ Glossary

Alloy: a metal made by melting two other metals together. Brass is an alloy of copper and zinc.

Anvil: a block of iron on which metals are worked in a forge.

Atom: a tiny particle of matter made up of electrons and a nucleus.

Barrel: oil used to be transported in barrels, and still is measured in barrels. One barrel is about 42 gallons (159 L).

Bellows: a simple machine that blows air into a forge or furnace to make the fire burn better.

Blacksmith: a craftsperson who makes things out of iron by heating the metal until it is soft enough to be shaped. A blacksmith who makes only horseshoes is called a farrier.

Carding: detangling wool or cotton by combing it. The first carding combs used were thistles (*carduus* is Latin for thistle). Today, machines with small metal spikes card the wool.

Chrysalis: the casing that forms around a caterpillar when it changes into a moth or butterfly.

Coke: the solid substance that is left behind after coal. Coke is a useful fuel.

Conductor: a substance through which a current, such as an electrical

current, can travel. Metals are good conductors of electricity. The electrical current in your house travels along copper wires. Metals are also good conductors of heat. A metal saucepan handle gets much hotter than a wooden one.

Copper: a reddish metal that is soft and easy to work and shape. Copper mixed with tin makes bronze. Before people discovered bronze, they used copper for objects that didn't have to be very strong. Look for ancient copper jewelry in museums.

Crucible: a container in which something is melted. In a blast furnace the crucible is the container at the bottom where molten metal collects. In a glassworks molten glass is left behind in the crucible.

Crystal: substances like salt, sugar, and some metals always form in the same geometric shapes—crystals. Rock crystal is a particular kind of stone, made of very clear crystals. Also, very clear high-quality glass is known as crystal. It makes a ringing sound when you tap it.

Deposit: an accumulation of a mineral, gas, or oil, which can be extracted from the ground.

Ductile: capable of being fashioned into a new form.

Energy: to cause a heat transfer between objects at different temperatures in order to create usable power.

Energy sources: electricity has to be made from some other kind of energy. Power stations where electricity is made use energy from oil, natural gas, water (hydroelectricity), coal, the heat of the sun (solar power), or wind power.

Extract: to take something out from the place where it was found. Miners extract coal from the ground. Other useful substances are extracted from metals and oil—for example, plastic is

extracted from the gases produced by heating oil.

Fibers: tiny threadlike strands that make up the leaves and stalks of plants. Glass fibers are threadlike strands of glass.

Fuel: any substance that is destroyed to produce energy. When you put wood on a fire, the fire destroys the wood and produces energy as heat. In the engine of a car, gasoline is destroyed by burning, but this produces the energy needed to operate the engine.

Furnace: a hot fire in an enclosed space. Bellows pump in the air that keeps the fire burning. Furnaces heat metals to high temperatures.

Grain: shows the pattern of the fibers in a material. The grain of a piece of wood can be seen on its surface when it has been smoothed or polished. Some types of wood have a distinctive and beautiful grain.

Hydrocarbons: substances made of carbon and hydrogen, such as oil and natural gas.

Impurity: another substance mixed in with a metal or precious stone, for example. Impurities in metal ores have to be removed before the metal can be used.

Insulate: to cover something that is a good conductor of heat or electricity with something that is a poor conductor. The rubber around wires in an electrical cord and the wooden handle on a saucepan act as insulators.

Kiln: an industrial oven. In a kiln, flames can't blacken an object being heated because the walls of the kiln protect it. Pottery and bricks are fired in a kiln.

Molten: melted by heat. Molten metal has been heated until it becomes liquid. A glassblower works with molten glass.

Mold: a hollow form into which a molten material can be poured or a soft material pressed. When the contents harden, they keep the shape of the mold.

Ore: a mineral from which a metal can be extracted by a process of heating it.

Recycle: to use something again. Paper can be recycled by mixing it with water to make pulp—which can in turn be made back into paper. Old glass jars and bottles can be melted down to make new bottles. Metal and plastic can be recycled too. Recycling helps preserve the Earth's resources, and slows the filling up of landfills.

Refine: to make something more pure. Gasoline is oil that has been refined.

Resin: the sticky sap of a tree. Synthetic resins can be manufactured to have the same qualities as sap.

Smelt: to heat an ore in order to extract a metal from it. Iron ore is smelted in a blast furnace, but aluminum ore has to be smelted at a much higher temperature.

Solder: a metal substance which when melted is used to join metallic surfaces.

Waste: what is left over when something has been extracted or manufactured. Some waste may contain dangerous chemicals and is difficult to dispose of safely. Most waste ends up in landfills, but can be burned or composted as well. Americans throw away 11 billion tons of solid waste a year.

■ **Have you heard these expressions?**

"She has a heart of stone."
(She is hard and cold.)

"He has a heart of gold."
(He is kind and good.)

"To rake someone over the coals."
(To speak to someone rudely or angrily)

"Strike while the iron is still hot."
(To act quickly at just the right moment)

"Every cloud has a silver lining."
(Problems or misfortunes often bring something good unexpectedly.)

"I slept like a log."
(I slept very heavily.)

"As thick as a brick"
(Very stupid)

"You can't make a silk purse out of a sow's ear."
(You can't make something fine and beautiful from the wrong materials.)

"Pull the wool over someone's eyes"
(To deceive someone by not letting them know the truth)

"Burn the midnight oil"
(Stay up very late)

These museums have fascinating exhibits on all sorts of subjects; many of them have hands-on activities. Check your local library for other interesting places to visit such as local industrial sites, a woodworking or glass blowing shop, or a recycling plant.

American Museum of Paper Making
500 10th Street NW
Atlanta, GA 30318
Telephone: 404-894-7840

Colorado School of Mines and Geology Museum
16th and Maple
Golden, CO 80401
Telephone: 303-273-3815

The Discovery Place
301 N. Tyron Street
Charlotte, NC 28202
Telephone: 704-372-6261

International Printing Museum
Buena Park, CA 90621
Telephone: 714-523-2070

Museum of Science and Industry
57th Street and Lake Shore Drive
Chicago, IL 60637
Telephone: 773-684-1414

National Ornamental Metal Museum
34 West California Avenue
Memphis, TN 38106
Telephone: 901-774-6380

The Science Museum of Minnesota
30 East 10th Street
St. Paul, MN 55101
Telephone: 612-221-9444

Texas Energy Museum
600 Main Street
Beaumont, TX 77701
Telephone: 409-833-5100

INDEX

The entries in **bold** refer to whole chapters on the subject.